1,000,000 Books

are available to read at

www.ForgottenBooks.com

Read online
Download PDF
Purchase in print

ISBN 978-1-5279-0167-4
PIBN 10929001

This book is a reproduction of an important historical work. Forgotten Books uses
state-of-the-art technology to digitally reconstruct the work, preserving the original format
whilst repairing imperfections present in the aged copy. In rare cases, an imperfection in
the original, such as a blemish or missing page, may be replicated in our edition. We do,
however, repair the vast majority of imperfections successfully; any imperfections that
remain are intentionally left to preserve the state of such historical works.

Forgotten Books is a registered trademark of FB &c Ltd.
Copyright © 2018 FB &c Ltd.
FB &c Ltd, Dalton House, 60 Windsor Avenue, London, SW19 2RR.
Company number 08720141. Registered in England and Wales.

For support please visit www.forgottenbooks.com

1 MONTH OF
FREE
READING

at
www.ForgottenBooks.com

By purchasing this book you are eligible for one month membership to ForgottenBooks.com, giving you unlimited access to our entire collection of over 1,000,000 titles via our web site and mobile apps.

To claim your free month visit:
www.forgottenbooks.com/free929001

* Offer is valid for 45 days from date of purchase. Terms and conditions apply.

English
Français
Deutsche
Italiano
Español
Português

www.forgottenbooks.com

Mythology Photography **Fiction**
Fishing Christianity **Art** Cooking
Essays Buddhism Freemasonry
Medicine **Biology** Music **Ancient
Egypt** Evolution Carpentry Physics
Dance Geology **Mathematics** Fitness
Shakespeare **Folklore** Yoga Marketing
Confidence Immortality Biographies
Poetry **Psychology** Witchcraft
Electronics Chemistry History **Law**
Accounting **Philosophy** Anthropology
Alchemy Drama Quantum Mechanics
Atheism Sexual Health **Ancient History**
Entrepreneurship Languages Sport
Paleontology Needlework Islam
Metaphysics Investment Archaeology
Parenting Statistics Criminology
Motivational

M2-U-5

University of Wisconsin Library

Manuscript Theses

Unpublished theses submitted for the Master's and
Doctor's degrees and deposited in the University of Wisconsin Library are open for inspection, but are to be used
only with due regard to the rights of the authors. Biblio-
graphical references may be noted, but passages may be copi
only with the permission of the authors, and proper credit
must be given in subsequent written or published work. Ex-
tensive copying or publication of the thesis in whole or in
part requires also the consent of the Dean of the Graduate
School of the University of Wisconsin.

This thesis by...................................
has been used by the following persons, whose signatures
attest their acceptance of the above restrictions.

A Library which borrows this thesis for use by its
patrons is expected to secure the signature of each user.

NAME AND ADDRESS   DATE

MAGNETIC PROPERTIES OF THE HEUSLER ALLOYS

by

HOWARD BOVEE BRIGGS

A Thesis Submitted for the Degree of

MASTER OF ARTS

UNIVERSITY OF WISCONSIN

1921

400259 AWM
NOV -1 1933 ·B 7672

The purpose of this work was to determine the magnetic
properties of the Heusler alloys, particularly with reference
to their behavior above the Curie point. The Heusler alloys,
as is well known, are magnetic alloys of copper, aluminum, and
manganese, which show remarkable magnetic properties. The
three constituents alone are each but slightly paramagnetic,
but when combined in certain proportions are decidedly ferromagnetic. The variation in magnetic strength with percentage
composition of each constituent was studied by several, among
them Knowlton [1], who came to the conclusion that the most
magnetic alloys were obtained when the combination was in the
proportion of the atomic weights of the constituents, ie.
Cu. 63.6, Al. 27.1, Mn. 54.9. He noted that the magnetic
strength at room temperature varied with the temperature from
which the alloy was quenched. Some interesting results in the
variation of the magnetic strength with the quenching temperature
were observed in the course of these experiments, and will be
discussed later in this paper.

The alloy was prepared according to the method suggested
by Knowlton [1]. Because of the affinity of manganese for
silicon, graphite crucibles and molds were used, and the melt
was made in a carbon resistor furnace. The proportions by
weight in the melt were Cu. 252 gr., Mn. 108 gr., Al. 40 gr.,
and the percentage composition was Cu. 63%, Mn. 27%, Al. 10%.
The copper was first melted and then the manganese added.
This was cast and remelted several times, after which the

(1)   (1) A.A.Knowlton, Physical Review 32, p. 54-68.

aluminum was added, and the charge cast and remelted several times again, in order to secure a good mixture. Finally the melt was cast in graphite molds into rods 1/4 in. in diameter and 4 in. long. The alloy was free from flaws, very hard, but it was found possible to turn specimens from it on the lathe.

The method of measurement of the susceptibility of the alloy is that of Curie, [1] with modifications introduced by Professor Terry. [2] The susceptibility is calculated from the mechanical force exerted on a specimen in a non-uniform field. For a small specimen placed between the poles of a magnet, the force urging it toward the stronger part of the field is

$$F = m \chi H \frac{dH}{dx}$$

where m is the mass of the specimen, χ the susceptibility referred to unit mass, and H the field strength. The force was measured on the Curie balance, a diagram of which is given in Fig. 1. The specimen in the form of a small cylinder was attached to the end of a porcelain rod which was supported by an aluminum arm. The whole was counterbalanced by a threaded lead nut W. The aluminum arm was rigidly attached to a movable coil B which was supported by a silk thread from a brass arm inside the case. The movable coil swung freely inside a fixed coil A. When the magnet was energized, the force on the specimen was counterbalanced by the electrodynamic action between the two coils. A concave mirror M was attached to the movable coil, and a spot of light focussed on a scale

(1) Curie, Oeuvres. p. 255, 1908.

(2) Terry, Earle M. Physical Review, p. 394.

four meters away indicated when a balance was obtained. The advantage of this form of instrument over that described by Curie lies in the ease of control, and the added range of forces which may be measured. Current was supplied to the movable coil by means of light coils of phosphor-bronze ribbon.

The furnace consisted of a tube of porcelain of 8 mm. internal diameter and 20 cm. long, closely wound with 0.5 mm. tungsten wire. Over this was placed another tube of porcelain, and the whole was mounted inside a water cooled brass jacket. Inside the furnace was placed a platinum platinum-rhodium thermocouple, protected by a porcelain sheath, the junction of the thermocouple extending to the middle of the furnace, and directly below the position to be occupied by the specimen. The tube containing the furnace and thermocouple was attached to the balance case by means of a ground joint. In calibrating the thermocouple the following melting points were used; Tin, 232°C., Zinc, 419°C., Antimony, 629°C. The melt specimens, in the form of small grains or strips, were placed in a small crucible of pure magnesia and lowered through the window L, until the crucible occupied the place that the specimen was to take later. A vacuum was then obtained. The melt specimen was illuminated by light reflected from a carbon arc by a prism into the furnace chamber, and observed by means of a tele-microscope. In this way the thermocouple was calibrated under the same conditions as later existed when it was used in the measurement of temperatures.

The magnet used was a large one, mounted in such a way that it could be moved forward or back by means of a slow

motion screw connected to the instrument table by means of a
rod. The pole pieces were 10 in. in diameter, and were given
the shape shown in Fig. 2 in order that the product

$$H \frac{dH}{dx}$$

should vary but slightly in the neighborhood of its maximum
value. This was desirable to minimize the error due to faulty
location of the specimen. As the specimen never exceeded 4 mm.
in diameter, and the measurements were made at the position of
maximum pull, the error due to faulty location of the specimen
was small.

The vacuum was obtained by means of a Cenco rotary oil
pump and a mercury diffusion pump, and read on a McLeod gauge.
Some difficulty was experienced in maintaining a good vacuum,
so the pump was run continuously during the measurements. The
vacuum was ordinarily better than .005 mm., and an examination
of the specimens after heating showed but slight oxidation.
A dish containing phosphorus pentoxide was placed in the case
near the opening to the furnace to absorb the water vapor.

In order to use the Curie balance to measure the
susceptibility of a specimen in absolute units, it was necessary
to determine the constant of the balance for a particular field
in terms of the currents in the fixed and movable coils. This
was done by determining the currents necessary to balance
specimens whose susceptibility had already been determined.
For this purpose, solutions of nickel sulphate, nickel nitrate,
cobalt sulphate, and cobalt nitrate, whose susceptibility had
been determined by the Quincke absolute method, were used.

The solution in a small glass vial was supported in the place
of the specimen, the normal field applied, and by shifting the
magnet back and forth, the position of maximum pull was located
and then the currents in the two coils recorded. The constant
of the balance was then obtained as:

$$F = m \chi H \frac{dH}{dx} = k_1 A_f A_m$$

$$k_1 = \frac{m \chi H \ dH/dx}{A_f A_m}$$

$$k = \frac{k_1}{H \ dH/dx} = \frac{m \chi}{A_f A_m}$$

$$\chi = \frac{k \ A_f A_f}{m}$$

Correction was made for the empty system, and by reversing the
currents in the balancing coils, the effect of the earth's
field was eliminated. The effect of the field of the magnet
on the coils was eliminated by a heavy shield made up of
several layers of rolled iron sheeting placed between the
magnet and the case containing the balance.

The specimen in the form of a small cylinder was attached
to the end of the porcelain support and the aluminum arm brought
into equilibrium in the center of the furnace by adjustment of
the lead nut W, and the phosphor-bronze ribbons. A vacuum was
then obtained. With the specimen in a highly magnetic state,
it was found impossible to balance the specimen, as the pull
was too great for the range of the balance, hence it was
necessary to heat the specimen to a temperature above its
inversion point before a balance could be obtained.

Summaries of the results obtained from three specimens, all of the same melt, are given below.

Table 1.

Specimen 1.

Two runs were made on this specimen. It melted off the support during the second run, at a temperature of $857^{\circ}C$.

Susceptibility as a function of the temperature.

Temp.	$\chi \cdot 10^{-6}$	Temp.	$\chi \cdot 10^{-6}$
444	7.91	25	14.33
425	7.60	40	14.03
404	7.36	80	12.66
380	7.18	160	11.38
356	7.07	205	10.09
---	6.93	240	8.41
298	7.00	244	8.25
275	7.18	270	7.53
257	7.36	298	7.07
240	8.00	328	6.85
201	9.46	363	6.73
180	10.30	412	7.13
112	12.72	460	8.20
15	14.48	508	9.15
		532	9.85
		575	9.34
		602	9.33
		630	9.24
		640	9.28
		675	9.02
		733	8.80
		805	8.45
		857	melt

The curves plotted from the above values show a decided minimum at 350°, a slight maximum at 515°, after which the susceptibility falls off gradually to the melting point. No maximum was observed in the region of the melting point, as was reported by Stevenson [1].

(1) Stevenson. E.B. Univ. Ill. Exp. St. Dec. 1910.

It is to be noted that the specimen on annealing from 444° does not regain its original state of magnetization. The values of the susceptibility on reheating are seen to correspond very closely to those on cooling.

Specimen 2.

Four runs were made on this specimen, of which the
first two are given below. The weight of this specimen was
0.0824 gr., or about one-tenth of the mass of the previous one.
It was thought that by taking a specimen of smaller mass a
balance at room temperature, with the alloy in its ferromagnetic state, would be possible. However, in spite of the
small mass of the specimen, a balance at room temperature was
only possible by reducing the field.

Susceptibility as a function of the temperature.

Temp.	$\chi \cdot 10^{-6}$	Temp.	$\chi \cdot 10^{-6}$
20°C.	2210	24°C.	38.40
40	2124	50	29.50
58	1985	63	27.05
82	1920	105	18.95
132	457	160	14.95
190	120	212	12.40
204	99	234	11.28
210	82.70	254	11.06
218	75.80	282	10.10
242	50.70	322	9.85
256	51.80	343	9.62
284	34.10	350	9.62
322	31.40	362	9.62
332	30.00	379	9.63
346	25.95	412	9.90
362	19.42	439	10.27
364	18.18	466	10.81
365	16.94	478	10.88
366	16.00	502	12.06
368	14.60	508	11.97
370	12.43	510	11.97
372	11.65	526	11.82
376	10.41	530	11.66
		550	11.50
		564	11.60
		588	11.43
		722	11.84
		724	11.84
		725	10.12

The first run on specimen #2 shows a slight decrease in susceptibility from room temperature to 100°, followed by a very rapid decrease from 100° to 200°, after which the susceptibility falls off more slowly. At 376° the specimen was allowed to cool off rapidly. As in the case of annealing from 444° with specimen 1, the alloy did not regain its original state, but increased slightly on cooling to $38.4 \cdot 10^{-6}$ at room temperature, or approximately 0.02 of its value in the highly magnetic state.

The curve of the second run on specimen 2 is seen to be very similar to the second run on specimen 1. Both were started in the weakly magnetic state, and show minima at 350° and maxima at about 515°.

At the end of the second run the specimen was allowed to cool off, and was followed with the balance to $92.0 \cdot 10^{-6}$, which was as strong a pull as the balance would handle. Apparently, then, the specimen on quenching from the region of its melting point regains its original highly magnetic state.

Table 3.

Specimen 2.

The following tables give the results of the third and fourth runs made on the second specimen.

Susceptibility as a function of the temperature.

Temp.	$\chi \cdot 10^{-6}$	Temp.	$\chi \cdot 10^{-6}$
710°C.	11.20	370	21.21
658	11.97	370.7	19.73
648	11.28	371.3	18.62
600	12.88	373	16.38
554	13.94	386.8	12.59
518	14.20	388	11.28
438	16.45	412	10.49
346	23.80	420	10.30
324	25.40	422	10.25
300	26.80	440	10.49
270	47.80	448	10.49
267	56.20	448	10.86
206	189.	470	11.38
190	525.	483	11.47
172	687.	513	12.12
172	687.	519	12.12
170	1077.	526	11.87
167	1376.	556	11.70
134	2590.		
126	2760.	Cooled and reheated	
126	2760.		
84	3980.	569	11.96
20	4130.	572	11.85
		598	11.74
		620	11.74
		636	11.64
		672	11.42
		710	11.07
		724	10.74
		764	10.42
		800	10.22
		717	11.27
		676	11.60
		633	12.57
		569	13.11

The third run on specimen two showed that the specimen, when annealed from the region of its melting point, regained its full magnetic strength. It appears that if the alloy is either quenched or annealed from a high temperature it regains its ferromagnetic state, while if it is either quenched or annealed from the region of its inversion point it is paramagnetic at room temperature. The above mentioned run does not show the minima at 350° and the slight maximum at 515°, so another run was made on the specimen. Run four shows a slight minimum at 422° and a maximum at 515°.

Table 4.

Specimen 3.

The third specimen used was hevier than the second, the purpose being to get more accurate balances and see to what degree Curie's law was obeyed.

Susceptibility as a function of the temperature.

Temp.	$\chi \cdot 10^{-6}$	Temp.	$\chi \cdot 10^{-6}$
542°C.	10.08	610	10.26
570	9.98	580	10.26
600	9.98	564	10.30
616	9.99	546	10.43
608	9.83	503	11.15
636	9.89	462	11.83
675	9.52	434	12.47
704	9.47	420	13.24
742	9.23	386	15.80
773	9.08	374	18.40
770	9.17	346	23.03
726	9.23	338	28.90
708	9.44	328	34.30
676	9.63	317	45.90
664	9.68	298	74.80

The results of the third specimen are the same as those found previously. In plotting the reciprocal of the susceptibility against the temperature it is at once evident that no straight line relation is followed.

Conclusion.

1. At room temperature the alloy may exist in two states, a paramagnetic state, and a ferromagnetic state.

2. The paramagnetic state is acquired when the alloy is either quenched or annealed from a temperature in the region of its transformation point.

3. The ferromagnetic state is acquired if the alloy is either quenched or annealed from a temperature in the region of its melting point.

4. Curie's law, ie. $\chi T = C$, is not obeyed for the region above the transformation point.

5. The inversion point for increasing temperatures is not the same as that for decreasing temperatures.

Printed by BoD™in Norderstedt, Germany